Roberto Souza

Geology of the Araticum Complex, Batalha-AL, Northeast Brazil

Roberto Souza

Geology of the Araticum Complex, Batalha-AL, Northeast Brazil

Southern domain of the Borborema Province. Geochemistry and economic geology of limestones for agricultural use produced in the region

ScienciaScripts

Imprint

Any brand names and product names mentioned in this book are subject to trademark, brand or patent protection and are trademarks or registered trademarks of their respective holders. The use of brand names, product names, common names, trade names, product descriptions etc. even without a particular marking in this work is in no way to be construed to mean that such names may be regarded as unrestricted in respect of trademark and brand protection legislation and could thus be used by anyone.

Cover image: www.ingimage.com

This book is a translation from the original published under ISBN 978-613-9-72642-4.

Publisher:
Sciencia Scripts
is a trademark of
Dodo Books Indian Ocean Ltd. and OmniScriptum S.R.L publishing group

120 High Road, East Finchley, London, N2 9ED, United Kingdom
Str. Armeneasca 28/1, office 1, Chisinau MD-2012, Republic of Moldova, Europe
Printed at: see last page
ISBN: 978-620-7-90233-0

Thank you

Thanks to God for the gift of life, my family, father, mother and sister for their constant support.

To Patrícia Rodrigues for her affection and support.

To LABISE and its ever-willing staff.

To the very dear Prof Alcides Sial.

To my supervisor in this work, Professor Valderez.

To FACEPE (process APQ1738-1.07/12), coordinated by Professor Valderez Pinto Ferreira, who financed the fieldwork that enabled this work to be carried out.

"If triangles had a God, he would have three sides"

Montesquieu

Summary

The Sergipana fold belt, geotectonically located between the south of the Pernambuco-Alagoas domain and the north-eastern limit of the São Francisco Craton, is made up of several tectono-stratigraphic domains: Estância Vaza Barris, Macururé, Marancó-Poço Redondo and Canindé domains. The Canindé Domain occurs as a band in a general NE-SW direction and is bounded from the other domains by large low-angle and high-angle shear zone faults. The study area, near the town of Batalha, Alagoas, comprises a strip of metasedimentary rocks called the Araticum complex, which is part of the Canindé domain. Field work indicates that the mapped region consists of garnet-biotite schists with intercalations of marbles and quartzites, garnet-biotite paragneisses with intercalations of amphibolites and marbles, these two main lithologies being separated by the Jacaré dos Homens shear zone (ZCJH). The schists and gneisses have foliation in the vicinity of the ZCJH with dips ranging from 15° to 70° and a general NE-SW direction. The lenses of metacalcaria intercalated in the schists and paragneisses occur as lenticular bodies, consistent with the regional structuring, elongated in a NE-SW direction, with dimensions varying from approximately 1 to 3 km in length and an average width of 450 metres. The mineral composition present (biotite, muscovite, garnet, plagioclase) in the schists and paragneisses indicates an amphibolite facies. The diagrams used to characterise the tectonic environment based on the chemical characteristics of the schists and paragneisses of the Araticum Complex suggest a passive continental margin environment. Major versus minor element contents (($\%$)Ni x ($\%$) TiO_2), (ppm)Rb x ($\%$)K_2O indicate an acidic to intermediate origin for these rocks. The CaO and MgO contents in the marble intercalations show that these rocks provide better chemical conditions for soils to develop agricultural crops.

SUMMARY

CHAPTER 1

Initial considerations

1.1   -Introduction

This work consists of geological mapping carried out in a region to the southwest of the municipality of Batalha, Alagoas, as part of the requirements for obtaining a bachelor's degree in Geology from the Federal University of Pernambuco.

In addition to geological mapping, this work aimed to characterise the lithology and geochemistry of the different lithologies found in the mapped area.

1.2   - Objectives

The main objective of this work was the geological mapping of an area of around 200 km^2 on a scale of 1:70,000 to the southwest of Batalha, Alagoas, and the geochemical characterisation of the metasedimentary rocks present, with the aim of inferring the provenance of the sediments and the tectonic environment of deposition. It also aimed to carry out a geochemical study in order to establish the possible applications in the agro-industry of limestones that occur in the mapped region.

1.3   - Location and access routes

The mapped area can be accessed from the city of Recife, Pernambuco on the BR-232 motorway, heading west. After the city of São Caetano, turn left towards the municipality of Garanhuns-PE, and follow the PE-218 motorway south to the city of Palmeira dos Índios-AL. From the city of Palmeira dos Índios, Alagoas, take the BR-316 motorway in a westerly direction and then continue to the city of Major Isidoro, Alagoas, in a southerly direction on the AL-120 motorway and then on to the city of Batalha, Alagoas, as shown in figure 1.1.

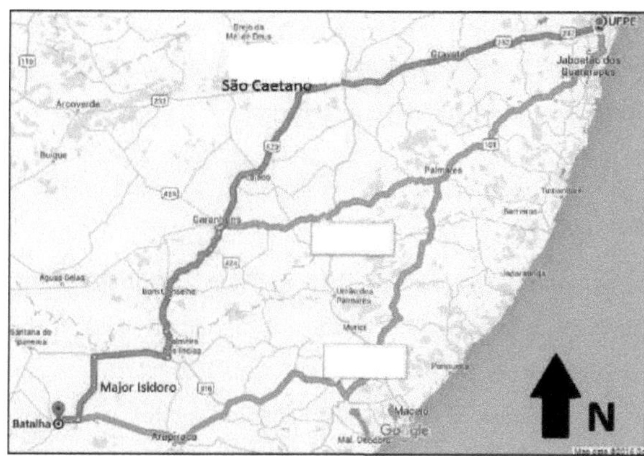

Figure 1.1 - Access map for the city of Batalha, Alagoas, from Recife, Pernambuco.

1.4 - Work methodology

The methodology of this work consisted of a field and laboratory phase.

These are: literature review, photointerpretation, field work, laboratory work and interpretation of the results. The first step consisted of gathering and reading works related to the geology and cartography of the region studied. Among these, the Pão de Açúcar topographic sheet (SC.24-X-D-IV), on a scale of 1:100,000, was used to delimit the study area, identify the road system, drainage and localities. In the next phase, aerial photographs on a scale of 1:70,000 were acquired from the library of the Geological Survey of Brazil (CPRM-Recife). These aerial photographs were used for stereoscopic geological photointerpretation (structures such as lineaments and foliations, for example), georeferencing and preparation of the geological service map for use in the field. During the field geological work, 31 outcrops were visited and samples collected. Geographical coordinates were obtained from each outcrop using a GARMIN GPS and, where present, geological structures were measured using a geologist's compass. The data acquired was then plotted on a map.

The material collected was taken to the sample preparation laboratory (LPA) at NEG/LABISE-UFPE (Geochemical Studies Centre and Stable Isotopes Laboratory) in the Geology Department (DGEO) and separated into three portions. One sample was crushed and pulverised and then sent to the DGEO's X-ray Fluorescence Laboratory to determine the concentrations of major elements and some trace elements. Another sample was sent to the Rolling Laboratory of the Department of Geology-UFPE to make thin slides for microscopic petrographic description. A third sample was preserved. Three samples were selected for separation of zircon grains for future U-Pb isotope analyses.

With the integration of field and laboratory data, geological maps and profiles were drawn up, along with chemical classification diagrams and a breakdown of possible tectonic environments. Finally, the results were interpreted and organised, culminating in this report.

1.5 - Physiographic aspects

The municipality of Batalha is physiographically part of the Diverse Dissected Surfaces unit, which occurs in the areas bordering the Piauí and Maranhão plateaus, in important areas of the Alagoas and Sergipe hinterlands and in small stretches of other states (IBGE, 1977).

The terrain is moderately dissected, with altitudes of between 300 and 700 metres, with poor, shallow soils, except in areas with narrow, deep valley bottoms (IBGE, 1977).

Surface water resources are very good due to the fact that the São Francisco River crosses the areas that make up this unit in the states of Alagoas and Sergipe (IBGE, 1977). The Ipanema River is the main landform and runs through the entire area of the city of Batalha, flowing into the São Francisco River.

The climate is characteristically very hot, with a rainy season in winter (~31 mm, SEMARH/AL). The rainy season begins in March and lasts until September (IBGE, 1977).

With regard to soils, on the tops of rounded reliefs and steep slopes there are litholic soils, which are shallow and stony and have medium natural fertility; on the lower slopes the soils are non-calcareous Bruno, with a clayey texture and high natural fertility; and on the flat tops there are latosols, which are deep, well-drained, acidic and have low natural fertility (IBGE, 1977).

CHAPTER 2

Regional Geology: Tectonic Compartmentalisation

The mapped area is part of the Borborema Province in northeastern Brazil, which is bounded to the east and north by the Atlantic Ocean and to the west by the Parnaíba Basin (Almeida et al., 1977) and consists of fold belts separated by extensive shear zones (Brito Neves et al., 1995), gneissic basement and igneous intrusions.

Brito Neves (1975) was a pioneer in recognising the main structural features of the Borborema province, recognising and naming the main domains (called fold belts and massifs) that are still used today (Figure 2.1), basically changing the names according to the author and the accumulation of new geological, geochemical and isotopic data.

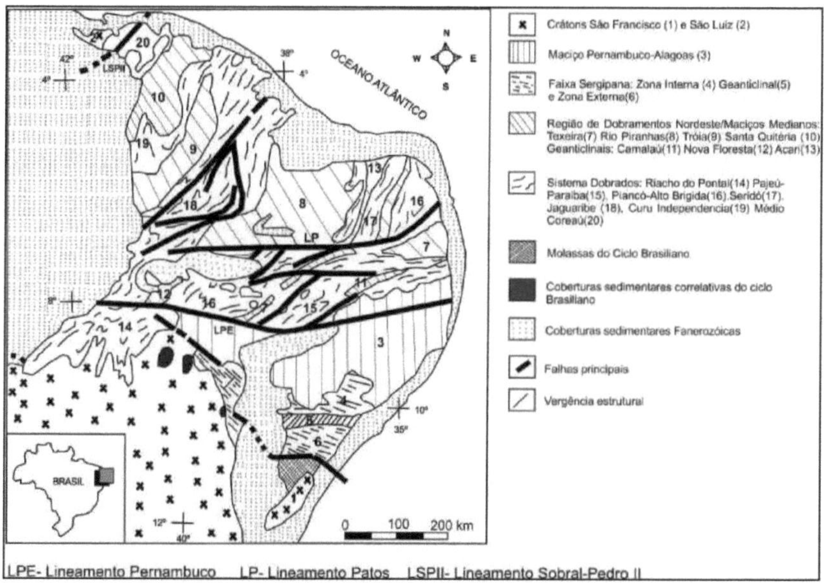

Figure 2.1 - Outline of the geological map of the Borborema Province, northeast Brazil (Brito Neves, 1975).

Santos and Brito Neves (1984) subdivided the territory of the Borborema Province based on structural domains, defining the Middle Coreaú, Far Northeast, Transnordestino or Central, Cearense and Sergipano domains.

Brito Neves et al. (2000) proposed another name for these five domains of the

Borborema Province, as proposed by Santos and Brito Neves (1984): Middle Coreaú, Central Ceará, Rio Grande do Norte, Central Domain (or Transverse Zone, named by Ebert, 1970) and Southern Domain (Figure 2.2).

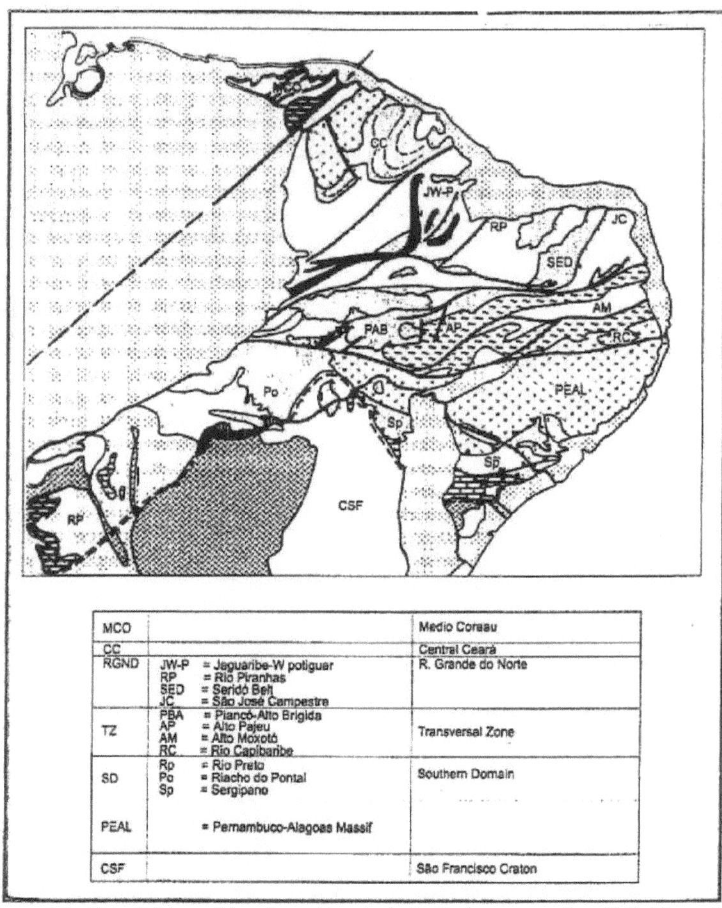

MCO			Medio Coreau
CC			Central Ceará
RGND	JW-P RP SED JC	= Jaguaribe-W potiguar = Rio Piranhas = Seridó Belt = São José Campestre	R. Grande do Norte
TZ	PBA AP AM RC	= Piancó-Alto Brigida = Alto Pajeu = Alto Moxotó = Rio Capibaribe	Transversal Zone
SD	Rp Po Sp	= Rio Preto = Riacho do Pontal = Sergipano	Southern Domain
PEAL		= Pernambuco-Alagoas Massif	
CSF			São Francisco Craton

Figure 2.2 - Main divisions of domains and terrains in the Borborema province (Brito Neves et al., 2000).

Van Schmus et al. (2011) regrouped these domains into three subprovinces in the Borborema Province: the northern subprovince, which encompasses the structural domains of Rio Grande do Norte and Ceará, north of the Patos shear zone and is made up of the Palaeoproterozoic basement (including Archaean cores) with overlying supracrustal rocks and Neoproterozoic plutonic rocks.

The central subprovince encompasses the domains that occur in the states of Paraíba and Pernambuco, between the Patos and Pernambuco shear zones, and is characterised by NE-SW transcurrent faults, already recognised by (Ebert, 1970) who called it the Transverse Zone.

The southern subprovince, as defined by Van Schmus et al. (2011) between the Pernambuco shear zone and the north of the São Francisco Craton, is made up of the Sergipano Domain, the Pernambuco-Alagoas Domain and the Riacho do Pontal Domain.

2.1 - Southern Domain

Brito Neves et al. (2000) subdivided the Southern Domain into four: the Rio Preto (1), Riacho do Pontal (2), Sergipano (3) and Pernambuco-Alagoas (4) fold belts. The Riacho do Pontal belt comprises metamorphosed sedimentary rocks (low to medium grade) pushed over gneisses from the São Francisco Craton (Jardim de Sá et al., 1992). The Rio Preto and Sergipano fold belts have Neoproterozoic supracrustal rocks at their southern end with similar stratigraphic sequences, typical of cratonic coverings. More distal sequences feature plutonic and volcano-sedimentary rocks (Brito Neves et al., 2000). The Pernambuco-Alagoas domain is a triangular region of ~70,000 km^2 that consists of granitic-migmatised basement with numerous Brasilian plutons (Brito Neves et al., 2000).

2.1.1 - Sergipano fold belt

The Sergipano fold belt is located between the north of the São Francisco Craton and the Pernambuco-Alagoas Domain. According to Davison and Santos (1989) and Silva Filho (1998) the Sergipano belt consists of six lithostratigraphic domains, each separated by a shear zone. These are the subdomains: Canindé, Poço Redondo, Marancó, Macururé, Vaza Barris and Estância. The last three are made up of metasedimentary rocks with a weak metamorphic grade in the Estância domain, passing through green shale in the Vaza Barris domain to amphibolite facies in the Macururé domain (Oliveira et al., 2006). The Canindé domain comprises a metavolcanic-sedimentary sequence represented by amphibolites, marbles, mica schists and metagrauvacas (Oliveira et al., 2006). The Poço Redondo domain is made up of migmatites, bitotite gneisses and various granitic intrusions (Oliveira et al., 2006). The Marancó domain contains a metavolcanic-sedimentary sequence (quartzites, conglomerates, mica schists, phyllites, andesite lenses) (Oliveira et al., 2006). The Macururé domain consists of granatiferous mica schists and phyllites with smaller amounts of quartzite and marble, all intruded by granitic bodies (Oliveira et al., 2006).

Silva Filho and Torres (2003) suggested two more domains: Rio Coruripe and Viçosa. The Rio Coruripe Domain was metamorphised into the granulite facies and then retrometamorphised into the amphibolite and green schist facies (Oliveira et al., 2006).

Mendes and Brito (2011) suggested a new tectonic compartmentalisation (Figure 2.3) for the Sergipano Fold Belt on the Arapiraca Sheet, where the part that was previously designated as the Rio Coruripe domain (Silva Filho, 2006) was designated as the Canindé domain (Figure 2.3).

The geotectonic compartmentalisation used in this work follows the configuration adopted by

Mendes and Brito (2011).

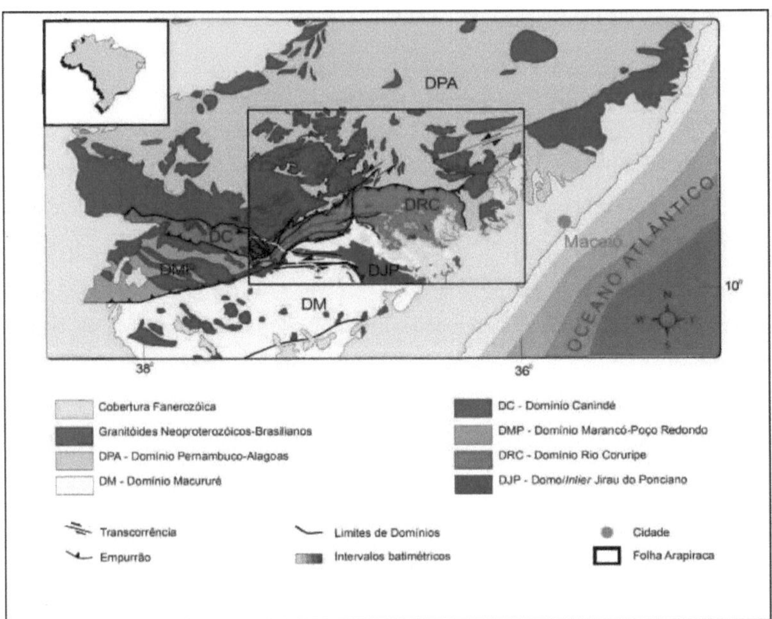

Figure 2.3 - Map showing the tectonic compartmentalisation of the Arapiraca sheet according to Mendes and Brito, 2011.

2.1.1.   1-Canindé Domain

The Canindé domain (Mendes et al., 2009) is made up of metavolcanic-sedimentary rocks from the Canindé Complex and amphibolites, marbles, mica schists and metagrauvacs from the Araticum Complex. The rocks of this subdomain show detrital zircons between 652 and 629 Ma, using the U-Pb method (Nascimento et al. 2005).

2.1.1.1- Araticum Complex

The Araticum Complex comprises a metavolcanic-sedimentary sequence made up of large (or not) biotite gneisses, biotite schists and metagrauvacs, with frequent intercalations of metamafic and metaultramafic rocks, marble lenses, graphitic schists, metamargas, calciosilicate rocks, banded ferriferous formations and even leucogranitoid syntectonic sheets (Mendes et al., 2009).

CHAPTER 3

Local Geology

The aim of this chapter is to describe the lithotypes identified in the study area, their structures and possible economic use. Descriptions of the outcrops were used to visualise the lithological framework, paying attention to structural criteria, contact relationships and superimposition of structures, as well as available bibliographical information.

The mapped region is covered by a thick mantle of soil and weathering and outcrops are scarce.

Several lithologies that form part of the Araticum Complex were identified in the area studied: garnet-biotite paragnase, garnet-biotite schist, amphibolite, marble and quartzite.

3.1 - Granada-biotite paragnaises

In the mapped area, the predominant lithology is greyish garnet-biotite gneiss with a grain size ranging from fine to medium. In hand samples, these rocks are composed of quartz (30-50%), biotite (20-40%), plagioclase (5-10%) and garnet (5-11%). Most of the outcrops visited are altered rocks (Figure 3.1 and Figure 3.2).

Figure 3.1 - Biotite gneiss cut with quartz-feldspathic vein. Outcrop RS-02

Figures 3.2 to 3.7 show rocks with compositional banding defined by quartz-feldspathic bands and biotite-rich bands, defining a very characteristic foliation.

Figura 3.2   Biotite gneiss showing marked foliation.
Attitude of the N2°W/60 foliation. Outcrop RS-03

Figura 3.3   - Garnet-biotite paragnase showing quartz-felsdspathic bands with biotite-rich bands, which are predominant. Foliation altitude N25°W/33°. Outcrop RS-04

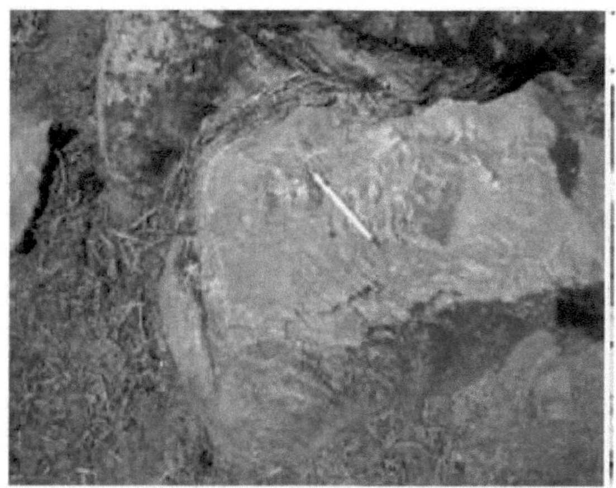

Figura 3.4   - Garnet-biotite paragnase showing folded quartz-felsdspathic bands. Foliation attitude N20°W/40°. Outcrop RS-06.

Figura 3.5   - Garnet-biotite paragnase, grey colour with biotite, quartz and garnet. Outcrop RS-09.

Figura 3.6   - Banded garnet-biotite paragnase with quartz-feldspathic bands and biotite-rich levels. Attitude of the foliation plane N20°W/40° Outcrop RS-10.

Figura 3.7   - Garnet-biotite paragnase with quartz-feldspathic bands interspersed with biotite-rich bands. Outcrop RS-17.

3.1.1 - Microscopic petrography

In thin section, these rocks have a granolepidoblastic texture with a modal composition: quartz (30 to 50%), biotite (30 to 40%), plagioclase (5-10%), garnet (10-20%), chlorite and zircon (less than 1%).

14

In the thin sections illustrated below (Fig. 3.8) quartz (40%) has a granoblastic texture, xenomorphic and showing undulating extinction. Garnet (12%) is fractured, xenoblastic. Biotite (35%) in this thin section occurs in a preferential direction defining a lepidoblastic texture and with the following pleochroism formula (Z=pale brown, X=light brown and Y=dark reddish brown). Zircon grains (less than 1%) are found embedded in biotite (Fig. 3.9)

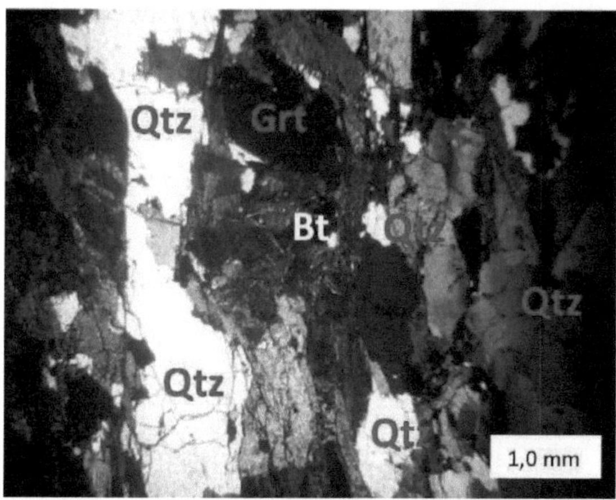

Figure 3.8 - Thin section of garnet-biotite paragnase seen through the petrographic microscope with crossed polarisers. Thin section RS-04 Qtz=Quartz, Grt=Garnet, Bt=Biotite.

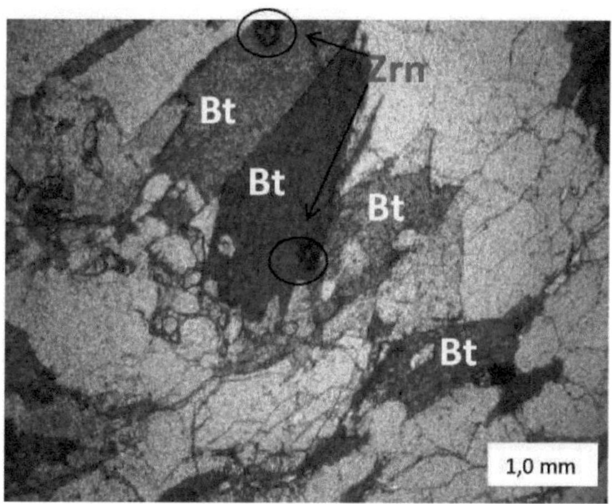

Figure 3.9 - Zircon grains (Zrn) embedded in biotite (Bt) in garnet-biotite paragnase. Thin section seen under the petrographic microscope with parallel polarisers. Thin section RS-04.

The thin section RS-10 (**Fig** 3.10) has the following composition: quartz (32%) with undulating extinction, garnet (10%), biotite (42%), plagioclase (15%) and zircon (less than 1%) included in the biotite (**Fig** 3.11).

Figura 3.10   - Thin section of garnet-biotite paragnase seen through the petrographic microscope with crossed polarisers. Qtz= Quartz, Pl = Plagioclase and Bt = Biotite. Thin section RS-10

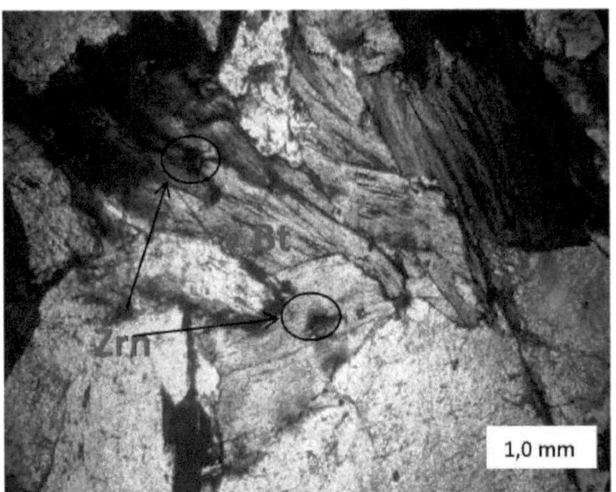

Figura 3.11   - Thin section of garnet-biotite paragnase seen through the petrographic microscope with parallel polarisers. Bt = Biotite, Zrn= Zircon Thin section RS-10

The thin section in figure 3.12 has a modal composition of quartz (40%), biotite (45%), sillimanite (2%), plagioclase (8%), garnet (5%) and opaque minerals (less than 1%). Sillimanite (fibrolite) is associated with biotite. Garnet is fractured, idioblastic with inclusions of biotite and opaque minerals (Fig. 3.13).

Figura 3.12   - Thin section of garnet-biotite paragnase seen through the petrographic microscope with parallel polarisers. Qzt= quartz, Bt = biotite, Grt= garnet ~1 mm long. Thin section RS-17

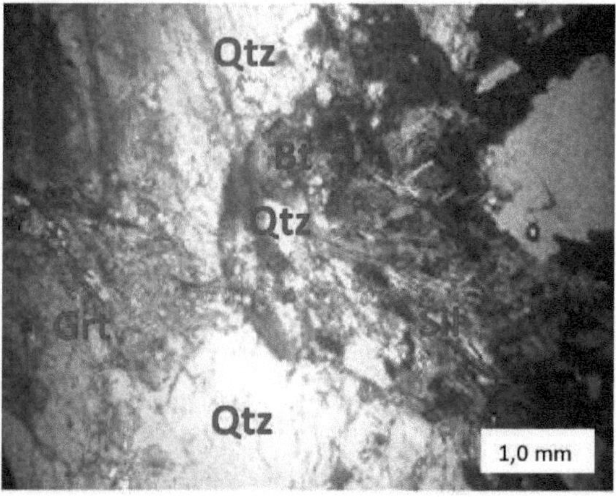

Figura 3.13   - Thin section of garnet-biotite paragnase seen under the petrographic microscope with parallel polarisers. Qtz=Quartz, Bt=Biotite, Grt=Garnet, Sill=Silimanite. Thin section RS-17.

The thin section of sample RS-06 shows schist made up of quartz (40%), biotite (35%) and plagioclase (25%) (**Fig.** 3.14) has a granolepidoblastic texture and shows xenoblastic quartz, with interlobed contacts and lepidoblastic biotite aligned along a foliation.

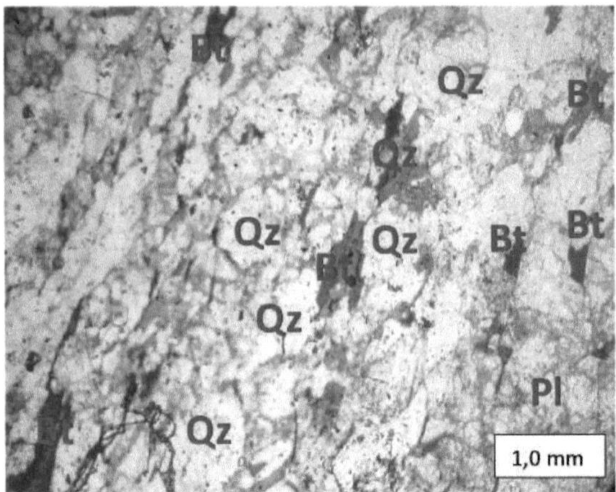

Figura 3.14 - Thin section of garnet-biotite paragnase seen through the petrographic microscope with parallel polarisers. Qtz=Quartz, Bt=Biotite, Grt=Garnet, Pl=Plagioclase. Thin section RS-06.

The thin section of sample RS-09 (**Fig.** 3.15) shows subidioblastic biotite (45%) with colours ranging from light brown to dark reddish brown (pleochroic formula: X=light brown, Y=pale brown and Z=dark reddish brown). The garnet is subidioblastic, fractured. Quartz is xenoblastic with undulating extinction. Opaque minerals occur next to the biotites.

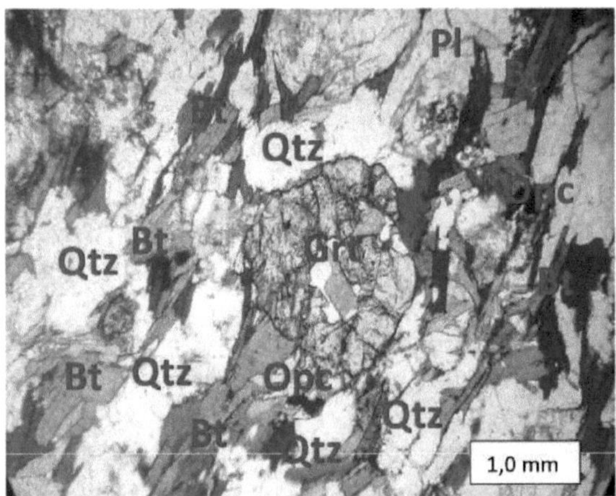

Figure 3.15 - Thin section of garnet-biotite paragnase seen through the petrographic microscope with parallel polarisers. Qtz=Quartz, Bt=Biotite, Grt=Garnet, Opc=Opacos. Thin section RS-09

3.2 - Garnet-biotite schists

This area is home to greyish garnet-biotite schists with a grain size ranging from fine to medium. In hand samples, these rocks show great compositional variation and can contain muscovite, biotite, megacrystals of garnet and quartz.

Figura 3.16 - Garnet profiroblasts in garnet-biotite schist. Outcrop RS-20.

Figura 3.17 - Garnet-biotite schist with marked foliation. Attitude of the foliation plane N10°E/50°. Outcrop RS-22.

Figura 3.18 - Garnet-biotite schist showing feldspar-rich bands showing pygmatic folds. Attitude of the foliation plane N6°W/33°. Outcrop RS-26

Figura 3.19 - Biotite schist with deformed garnet porphyroblasts, showing well-developed pressure shadow. Outcrop RS-11.

Figura 3.20   - Garnet-biotite schist. Medium granulometry (2 mm) given by garnet porphyroblasts
throughout the rock. Outcrop RS-13

Figura 3.21   - Garnet-biotite schist. The biotite defines a well-developed foliation. Outcrop RS-27.

3.2.1- Microscopic petrography

The thin section of sample RS-12 (Fig. 3.22) shows a granolepidoblastic texture with quartz (35%), biotite (44%), muscovite (5%), plagioclase (10%) and garnet (16%).

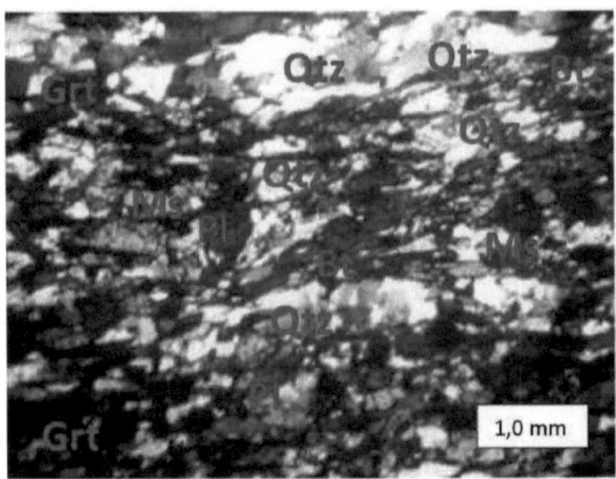

Figura 3.22 - Thin section of garnet-biotite schist showing granolepidoblastic texture under crossed polarisers. Qtz=quartz, Bt=biotite, Ms=muscovite, Grt=granate Thin section RS-12

The thin section of sample RS-11 (Fig. 3.22) shows quartz (22%), muscovite (25%), biotite (15%), plagioclase (17%), garnet (20%) and chlorite (1%). Muscovite has a lepidoblastic texture, quartz is xenoblastic with undulating extinction, garnet is fractured when observed under natural light with parallel polarisers.

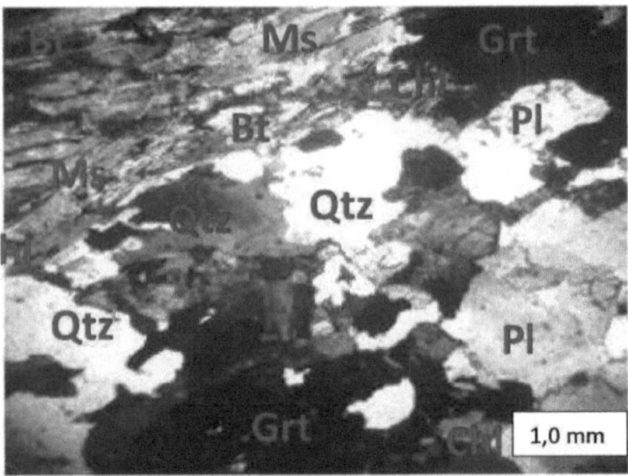

Figura 3.23 - Thin section of garnet-biotite schist under a petrographic microscope with crossed polarisers. Qtz=quartz, Bt=biotite, Ms=muscovite, Grt=garnet, Pl=plagioclase and Chl=chlorite. Thin section RS-11

The thin section RS-13 (Fig. 3.24) has quartz (35%), biotite (42%), plagioclase (7%) and garnet (15%). The biotite gives it a lepidoblastic texture, the garnet shows fractures and quartz inclusions (Fig.3.24). Zircon inclusions can be seen in the biotite (Fig. 3.25).

Figura 3.24   - Thin section of garnet-biotite schist seen through the petrographic microscope with crossed polarisers. Qtz=quartz, Bt=biotite, Grt=garnet, Pl=plagioclase. Thin section RS-13.

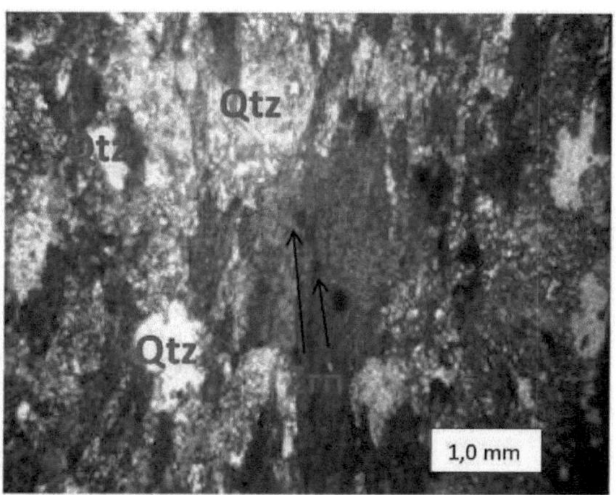

Figura 3.25   - Thin section of garnet-biotite schist under a petrographic microscope with parallel polarisers. Qtz=quartz, Bt=biotite, Ms=muscovite, Grt=garnet, Zrn=Zircon. Thin section RS-13

The thin section of sample RS-22 (Fig. 3.26) has a granoblastic texture represented by xenoblastic grains of quartz (45%) with undulating extinction, palettes of biotite (27%) giving the blade a lepidoblastic texture, plagioclase (8%), garnet (20%).

Figure 3.26 - Thin section of garnet-biotite schist in a petrographic microscope with crossed polarisers. Qtz=quartz, Bt=biotite, Grt=garnet, Pl=plagioclase Thin section RS-22.

The thin section of sample RS-20 (Fig. 3.27) shows quartz (34%), muscovite (28%), plagioclase (8%), muscovite (42%), garnet (30%). The muscovites are lepidoblastic, the quartz appears as xenoblastic grains, the garnets are very fractured when viewed under natural light with parallel polarisers.

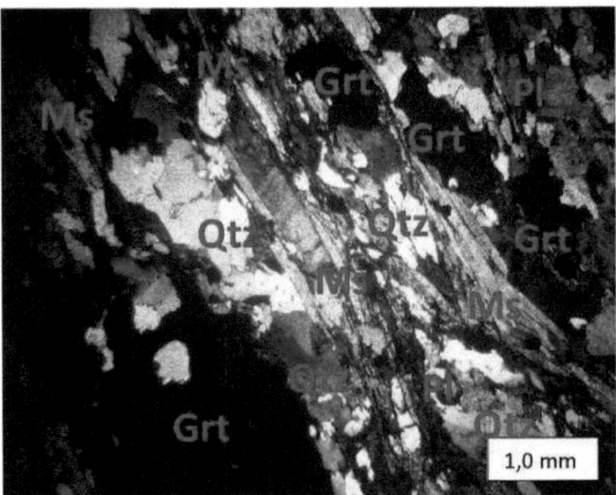

Figure 3.27 - Thin section of garnet-biotite schist in a petrographic microscope with crossed polarisers. Qtz=quartz, Ms=Muscovite, Grt=garnet, Kfs=potassiumfeldspar. Thin section RS-20.

A well-developed granolepidoblastic texture is visible in the thin section of sample RS-26 (Fig. 3.28) represented by palettes of biotite (48%), quartz (35%) and plagioclase (8%). The quartz is presented in xenoblastic grains with undulating extinction, while the biotite occurs in the lamina in palettes scattered throughout the thin section, giving it a lepidoblastic texture.

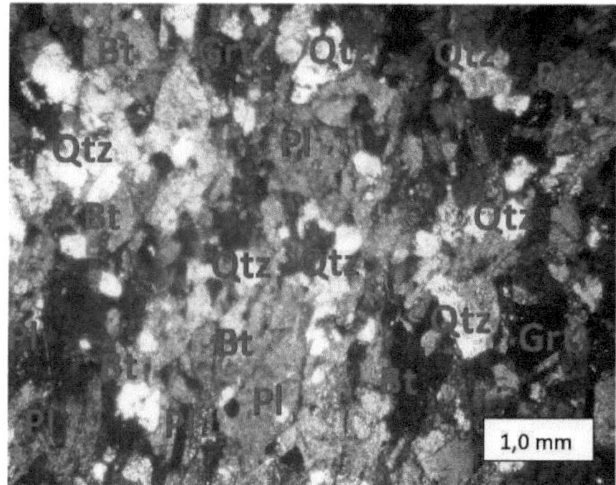

Figure 3.28 - Thin section of garnet-biotite schist under a petrographic microscope with crossed polarisers. Qtz=quartz, Ms=Muscovite, Grt=garnet, Pl=Plagioclase.Thin section RS-26

The thin section of sample RS-27 (Fig. 3.29) is composed of quartz (41%), biotite (40%), plagioclase (14%) and garnet (5%). This rock has a granolepidoblastic texture defined by the biotite disseminated throughout the sheet and the quartz grains.

Figure 3.29 - Thin section of garnet-biotite schist in a petrographic microscope with crossed polarisers. Qtz=quartz, Ms=Muscovite, Grt=garnet, Pl=Plagioclase. Thin section RS-27

3.3 - Amphibolites

These rocks are distributed in the northern part of the mapped area, forming 6 lenses of up to . They are dark grey-green in colour and show well-developed foliation following a SW-NE direction, preferentially parallel to the regional foliation.

25

Figura 3.30- - Dark greenish-grey amphibolite with a folded leucocratic band. Rock composed of amphibole, plagioclase and biotite. Foliation altitude N33°W/40°. Formation RS-05.

Figura 3.31- Dark greenish-grey amphibolite with a mineralogical composition of amphibole and plagioclase. Foliation altitude N33°W/40°. Outcrop RS-08.

Figura 3.32-   Dark greenish-grey amphibolite with fine to medium grain size. This rock has bands of plagioclase interspersed with levels of amphibole and biotite that define a slightly crenulated foliation. Attitude of the foliation plane N25°W/35°. Outcrop RS-18.

Figura 3.33-   Dark greenish-grey amphibolite composed of plagioclase and amphibole. The main foliation is cut by fractures spaced ~10 cm apart. Attitude of the foliation plane N20°W/35°. Outcrop RS-16.

3.3.1 - Microscopic petrography

In thin section, sample RS-15 (Fig. 3.34) has a granoblastic texture and is made up of hornblende (50%), plagioclase (40%) and titanite (10%). The hornblende is idioblastic and shows titanite on its edges.

27

Figure 3.34 - Amphibolite with horblende-Hbl, plagioclase-Pl and titatin-Tn on the edges of the hornblende. Observed in the petrographic microscope with parallel polarisers. The largest hornblende crystal is ~1.00 mm. Thin section RS-15.

The thin section of sample RS-05 (Fig. 3.35) has a granoblastic texture made up of hornblende (55%) disseminated throughout the blade, plagioclase (45%) and titanite (less than 1%).

Figura 3.35- Amphibolite with horblende-Hbl, plagioclase-Pl. Observed in the petrographic microscope with parallel polarisers. The largest hornblende crystal is ~1.00 mm. Thin section RS-05.

The thin section of sample RS-08 (Fig. 3.36) has a granoblastic texture with subdiomorphic crystals of hornblende (55%) and plagioclase (45%).

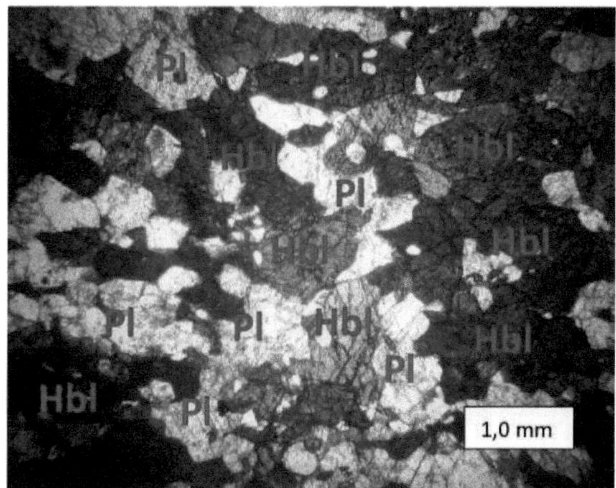

Figura 3.36-   Amphibolite observed in the petrographic microscope with parallel polarisers. The largest hornblende crystal is ~1.00 mm. Horblende-Hbl, plagioclase-Pl. Thin section RS-08.

The thin section of sample RS-16 (Fig. 3.37) has a granoblastic texture with the presence of horblende (55%), plagioclase (44%) and titatine (less than 1%).

Figura 3.37-   Amphibolite observed in the petrographic microscope with parallel polarisers. The largest hornblende crystal is ~1.00 mm. Horblende-Hbl, plagioclase-Pl, titanite-Tn. Thin section RS-16

This thin section (RS-14) (Fig. 3.38) has a granoblastic texture with hornblende (55%) and plagioclase (45%) xenoblastic minerals.

Figure 3.38 - Amphibolite observed in the petrographic microscope with parallel polarisers. The largest hornblende crystal is ~1.00 mm. Horblende-Hbl and plagioclase-Pl. Thin section RS-14.

3.2   - Metacalcarians

The metacalcareous rocks found in the study area are lenticular in shape, cutting through the paragneisses and schists and conforming to the regional structures with a longitudinal extension of between 1 and 3 kilometres. These rocks have a massive texture and a beige to white colour.

Figura 3.39   - Active mine exploiting metallurgical limestone for agricultural purposes.
Site: Incorpore Corretivos Agrícolas Ltda. Outcrop RS-01.

Figura 3.40   Slightly altered, beige-coloured metacalcareous outcrop.
Outcrop RS-19

Figure 3.41 - Metacalestone with a massive beige texture.
Outcrop RS-19.

Figure 3.42 - White-coloured massive metacalcareous rock at outcrop RS-23.

Figura 3.43- Metacalcareous rock with a massive texture and beige colour from the RS-23 outcrop. Scale graduated in centimetres.

Figura 3.44- Slightly foliated metasedimentary with granoblastic texture in outcrop RS-25

Figure 3.45. Metacalcareous rock with a massive texture, beige colour, from outcrop RS-25. Scale graduated in centimetres

3.4 - Quartzites

The quartzite present in the area is pink in colour and medium to coarse-grained. They are essentially composed of quartz and various levels of muscovite. They show a very marked foliation defined by stretched quartz grains and muscovite that occurs aligned on the foliation planes (Fig. 3.47 and Fig. 3.49). In thin section, the quartz grains show strong undulating extinction. The muscovite flakes have a crystallographic orientation (all the grains die out at the same time), according to the foliation.

Figura 3.46   Rose-coloured quartzite, medium to coarse granulometry in the RS-21 outcrop.

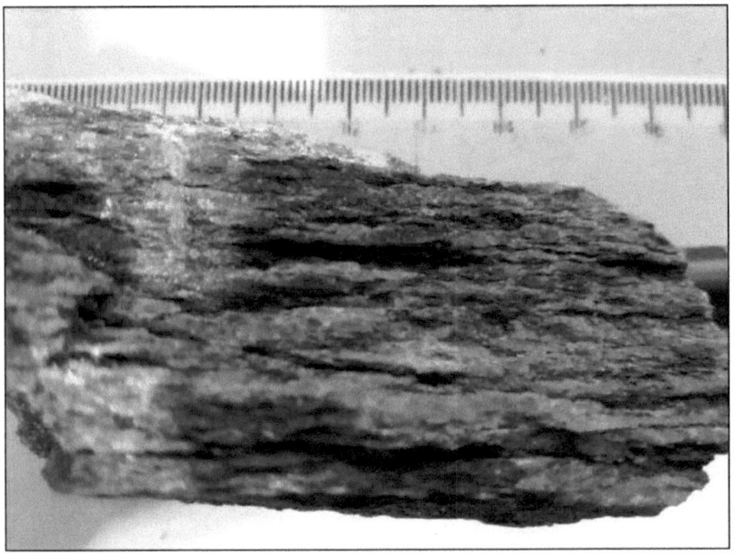

Figura 3.47   - Quartzite showing well-dissolved foliation marked by the stretching of the quartz grains (outcrop RS-21). Scale indicated in centimetres.

Figura 3.48 - Rose-coloured quartzite, medium to coarse granulometry in the RS-24 outcrop.

Figura 3.49 - Quartzite showing stretched quartz in the sample from outcrop RS-24. Scale indicated in centimetres.

CHAPTER 4

Structural geology

The structural study of the geological units in the mapped area began with the analysis of structural features on a macroscopic scale, using aerial photographs, and on a mega/mesoscopic scale through observations and measurements of the attitudes of features such as foliation in the outcrops found in the field. According to Billings (1965), foliation is a property of igneous or metamorphic rocks where they break along parallel surfaces, defined by placer or prismatic minerals. Schistosity is defined by the preferential orientation of micaceous minerals along parallel or subparallel surfaces, according to which the rock can break into more or less irregular slices or sheets. Throughout the area mapped here, the schists and paragneisses show foliation and banding defined by minerals in a planar arrangement such as biotite, muscovite, quartz-feldspathic levels defining a characteristic foliation which allowed the attitude of these planes to be gauged. In the area mapped, the 14 foliation poles measured in these rocks show a preferential NNW-SSE trend, with dip intensity varying between 15° and 70°. As the stereogram in Figure 4.1 shows, the concentration of poles plotted between the second and third quadrants confirms the preferential direction of dips to the NNW. Confirming the above measurements, according to Brito et al. (2007) the Canindé Domain, in the Batalha region, Alagoas, represented by the metasedimentary rocks of the Araticum Complex, is distributed in an elongated band in a NE-SW direction.

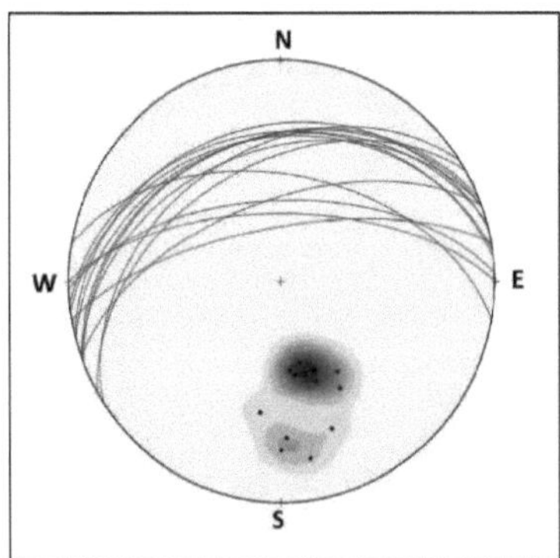

Figure 4.1 - Stereogram of the poles of the schist and paragneiss foliations, indicating a preferential NW-SE dip and a preferential NE-SW direction.

CHAPTER 5

Economic geology

The metacalcareous samples collected in the study area were taken for chemical analyses and a study to assess their application in agriculture. To do this, the equation for calculating the neutralising power was used to gauge its ability to neutralise soil acidity for agricultural purposes.

The neutralising power (NP) is the capacity of a limestone to neutralise the acidity of the soil (CBPM, 1998). It is determined by the ratio between the limestone and an acid, but can be obtained by calculating the chemical composition of the limestone, considering the equivalent weight of $CaCO3 = 100\%$ (CBPM, 1998). The calculation of calcium carbonate equivalence is made by converting the CaO and MgO contents (it can also be done from CaCO3 and MgO) into equivalent values (Eq.) of CaCO3, using the equation **(% CaO x 1.79) + (% MgO x 2.48) = % Eq. CaCO3** (CBPM, 1998).

The result of calculating the neutralising power of the limestones studied using the equation published by the CBPM above is shown in table 5.1. These rocks show a neutralisation power of over 90%, which indicates excellent results for reducing the acidity of soils for agricultural purposes.

Sample	Calculated Neutralising Power
RS-19	100,62
RS-23	108,32
RS-01	103,89
RS-25	104,05

Table 5.1 - Neutralising power calculated from the MgO and CaO contents of four samples.

The Agricultural Inspection Secretariat (SEFIS) published in the Federal Official Gazette on 16.06.86 Ordinance No. 3, which establishes MgO levels for limestone for use in agriculture, and which was complemented by the Bahia State Soil Fertility Commission, which adopts CaO concentration ranges, as shown in Table 5.2 below:

Limestone	CaO content (%)	MgO content (%)
Calcitic	40-45	<5
Magnesian	31-39	5 -12
Dolomitic	25-30	>12

Table 5.2- Classification of agricultural limestones (BRASIL. Ordinance No. 3, 16/06/1986).

According to this classification, samples RS-01, RS-23 and RS-25 can be classified as magnesian to calcitic limestones, as they have CaO contents between 33.22% and 40.94%. According to the MgO content, these samples are dolomitic because they have MgO contents greater than 12%. Sample RS-19 showed a MgO content of less than 3.51% and a CaO content of more than 40% (51.35%), making it a calcitic limestone.

Samples RS-01, RS-23 and RS-25 can be classified as magnesian to dolomitic limestones, as they have CaO contents between 33.22% and 40.94%, while their MgO contents are higher than 12%. Sample RS-19 showed a MgO content of less than 3.51% and a CaO content of more than 40% (51.35%).

CHAPTER 6

Geochemistry

The geochemistry of metasedimentary rocks was used to characterise the mapped paragneisses and schists shown in this work. Clastic metasedimentary rocks can have their probable protoliths inferred using their chemical composition and comparing it with the typical composition of sedimentary rocks, assuming closed metamorphism, i.e. there was no addition or loss of matter, only heat exchange. Similarly, the provenance of the clasts in siliciclastic sedimentary rocks and the tectonic environment in which the sediments were deposited can be inferred using the chemistry of major and trace elements in the whole rock.Herron (1988) and Pettjohn (1972) proposed some chemical classifications based on the contents of major elements in the whole rock, which are illustrated in Figures 6.1 and 6.2 respectively.Taking the relationships between SiO2/Al2O3 vs Fe2O3/K2O and Na2O/K2O in the diagram of the chemical classification of sedimentary rocks by Herron (1988) illustrated in figure 6.1 and Pettijohn (1972) illustrated in figure 6.2, it can be seen that the protoliths of the metasedimentary rocks of the Araticum Complex analysed here correspond to shales, sandstones and grauvacas.

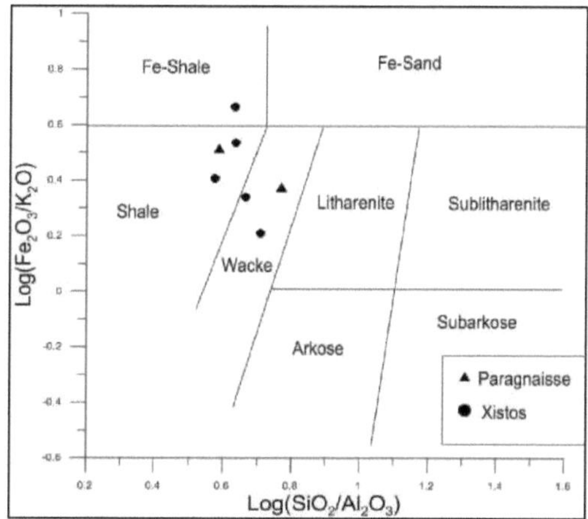

Figure 6.1 - Discriminant classification diagram (according to Herron, 1988) of siliciclastic sediments by their logarithmic SiO2/Al2O3 and Fe2O3/K2O ratios.

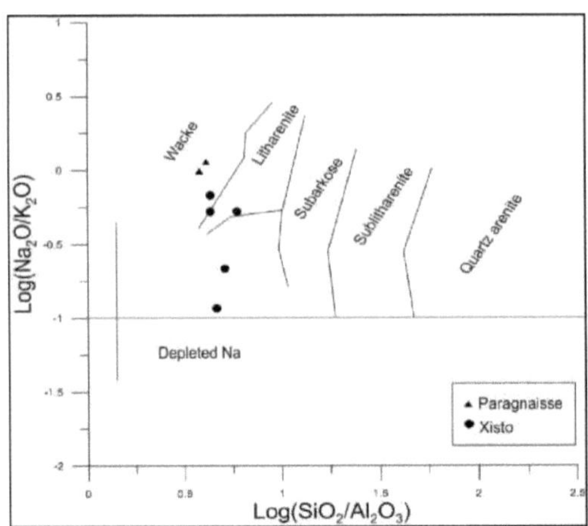

Figure 6.2 - Discriminant classification diagram (according to Pettjohn, 1972) of siliciclastic sediments by their logarithmic SiO2/Al2O3 and Na2O/K2O ratios.

The K2O versus Rb diagram of Floyd et al., 1989 was proposed to distinguish the igneous source from the clasts of siliciclastic rocks. The chemical compositions of the samples shown here in figure 6.3 are all plotted in the field of igneous source materials with an intermediate+acid composition. The TiO2(%) versus Ni(ppm) diagram proposed by Floyd et al (1989) applied to the samples studied (figure 6.4) indicates that the sediments of the (meta)sedimentary samples studied come from acidic to intermediate igneous rocks, consistent with what is indicated in the diagram by Floyd and Leveride (1978).

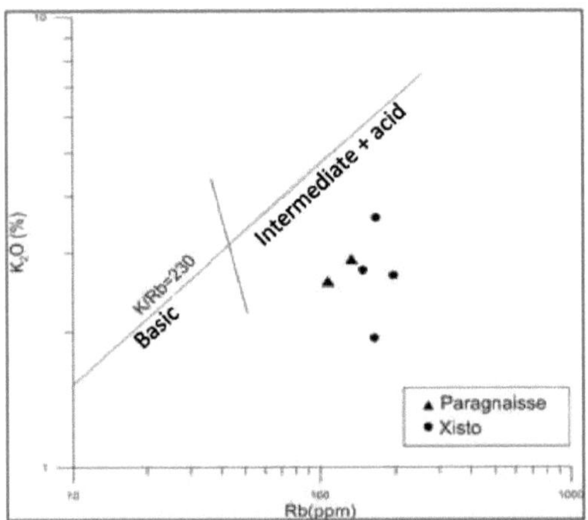

Figura 6.3 - K-Rb diagram (Floyd and Leveridge, 1987) to discriminate between (meta) sedimentary rock sources of magmatic origin with basic and intermediate+ acid compositions.

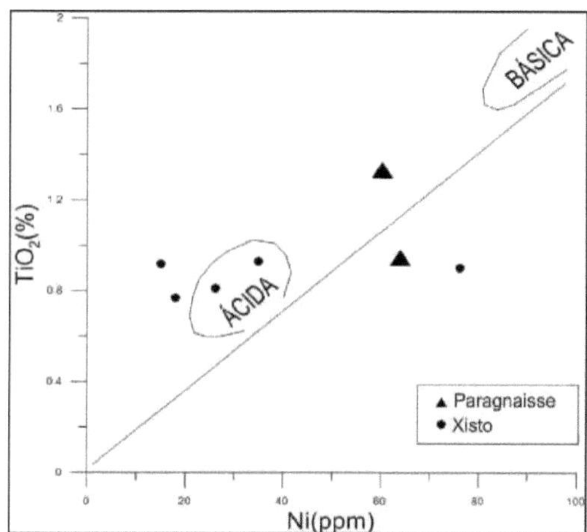

Figura 6.4 - TiO2 versus Ni for the metasedimentary rocks of the Araticum Complex studied here. The fields indicate the acidic to intermediate composition, according to Floyd et al., (1989) of the protoliths of the rocks studied.

The Bhatia diagram (1983), which is used to discriminate tectonic environments based on the chemical characteristics of sedimentary rocks, suggests that the metasediments of the Araticum Complex would be related to an active continental margin environment. The graphs plotted in these diagrams ratify the study by Oliveira et al., 2010 on the Sergipano Belt and Canindé Domain.

Oliveira et al. (2010) proposed that the evolution of the Sergipano Belt began with the break-up of the Palaeoproterozoic continent followed by the development of the Mesoproterozoic continental arc (Poço Redondo Gneisses) possibly on the margin of the Pernambuco-Alagoas Block (Oliveira et al., 2010). The extension of this continental block gave rise to (i) A-type granites associated with sedimentary rocks of the stretched margin of the Poço Redondo-Marancó Domain, (ii) between the Pernambuco-Alagoas Domain and the Poço Redondo/Marancó Domain the Canindé volcano-sedimentary sequence, (iii) and a passive margin at the southern end of the Pernambuco-Alagoas Block.

Sedimentation in the Canindé Domain probably began around 715 M.y. (U-Pb granite Garrote type A age) and continued until at least 625 M.y. (the age of the detrital zircons of the Novo Gosto-Mulungu unit) (Oliveira et al., 2010).

The convergence of the Pernambuco-Alagoas Block and the São Francisco Craton led to the deformation of sediments and the construction of a continental arc between 630 M.y. and 620 M.y. in the Macururé, Poço Redondo-Marancó and Canindé domains (Oliveira et al., 2010).

41

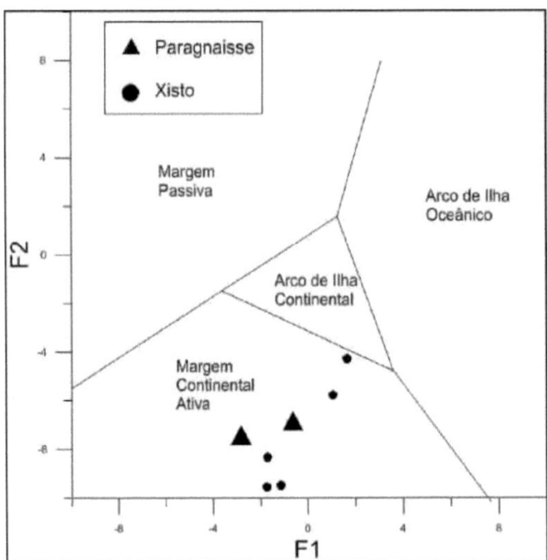

Figure 6.5 - Diagram of discriminant functions F1 and F2 for sedimentary rocks (Bathia, 1983): F1 = -0.0447 x (SiO_2) - 0.972 x (TiO_2) + 0.008 x $(Al\ O_{23})$ - 0.267 x $(Fe\ O_{23})$ + 0.208 x (FeO) - 3.082 x (MnO) + 0.140 x (MgO) + 0.195 x (CaO) + 0.719 x (Na2O) - 0.032 x (K2O) + 7.510 x (P2O5) + 0,303 and F2 = -0.421 x (SiO2) + 1.988 x (TiO2) - (0.526 x Al2O3) - 0.551 x (Fe2O3) - 1.610 x (FeO) + 2.720 x (MnO) + 0.881 x (MgO) - 0.907 x (CaO) - 0.177 x (Na2O) - 1.840 x (K2O) + 7.244 x (P2O5) + 43.57.

CHAPTER 7

Conclusions

The area studied is mainly made up of biotite and garnet-rich paragneisses, containing lenses of amphiboles and metacalcites, and biotite schists with garnet containing lenses of metacalcites and quartzites.

The mineral assemblage: biotite, muscovite, garnet, quartz and plagioclase in the schists and paragneisses described in the study area are indicative of progressive metamorphism in the amphibolite facies. A retrometamorphic phase is suggested by the presence of chlorite, light green in colour under parallel polarisers, lepidoblastic and found in thin section near the biotite.

The petrographic and geochemical studies carried out allow us to conclude that the protoliths of the clasts of the metasedimentary rocks of the Araticum Complex mapped in this area can be characterised as belonging to a sequence of clastic sedimentary rocks formed by sandstones, grauvacas and shale, from acid to intermediate rocks, deposited on the passive continental margin.

The chemical compositions of metacalcareous rocks indicate that they provide the best chemical conditions for use in soils for the purpose of growing agricultural crops.

Bibliographical references

Almeida, F. F. M.; Hasui, Y.; Brito Neves, B. B. 1977. The Upper Precambrian of South America. Boletim ig/usp. V.7, p. 45-80.

Bhatia, M.R., 1983. Plate tectonics and geochemical composition of sandstones: Journal of Geology, 91(6), 611627.

Billings, M. P. Structural Geology - 1 ed. - New York : Prentice Hall of New York, 1965. Page 213

Brito Neves B.B., Van Schmus W.R., Santos E.J., Campos Neto N.C., Kozuch M, 1995. The Cariris Velhos Event in the Borborema Province: Data integration, implications and perspectives. Revista brasileira de geociências, 25(4):279- 296.

Brito Neves, B. B.; Santos, E. J.; Van Schmus, W. R. 2000. Tectonic History of the Borborema Province. 31st International Geological Congress, Rio de Janeiro, Brazil. Pages 151-155

Brito Neves, B. B. de. 1975. Geotectonic regionalisation of the Northeastern Precambrian. Thesis - Institute of Geosciences, University of São Paulo, São Paulo. Page 21

Brito, M. F. L.; Mendes, V. A.; Paiva, I.P.; Wanderley A. A.; Medeiros, V. C. Field and Petrographic Aspects of the Metasedimentary Rocks of the Araticum Complex, Canindé-Marancó Domain, Sergipano Fold System. XXII Northeast Geology Symposium. 2007

Bahia Mineral Research Company. Agricultural Limestone: Diagnosis of Supply and Demand in the State of Bahia. Salvador : EBDA : CBPM. Pages 12-18 (documents, 8). 1998

Davison, I., Santos, R.A., 1989. Tectonic evolution of the Sergipano fold belt, NE Brazil, during the Brasiliano orogeny. Precambrian Research. 45, 319-342

Ebert, H. The Precambrian geology of the "Borborema" Belt. 1970 State of Paraíba and Rio Grande do Norte, northeastern Brazil. Geol. Rundsch. V.59, n.3, p.1299-1326.

Floyd, P.A., Leveridge, B. E., (1987). Tectonic environment of the Devonian Gramscatho basin, south Cornwall: framework mode and geochemical evidence from turbidite sandstones: Journal of the Geological Society of London, 144. 531-542.

Floyd, P. A., Winchester, J. A., Park, R. G., 1989. Geochemistry and tectonics of lewisian clastic metasediments from the early proterozoic lock marie group of gairlock. Scotland: Pre-cambrian Reserach 45. 203-214.

Brazilian Institute of Geography and Statistics - IBGE. Geography of Brazil. Northeast Region. Rio de Janeiro: SERGRAF. Available on 1 cd. 1977

Herron, M.M., 1988. Geochemical classification of terrigenous sands and shales from core or log data: Journal of Sedimentary Petrology 58. 820-829.

Jardim de Sá E.F., Macedo M.H.F., Fuck R.A., et al., 1992. Proterozoic terrains in the Borborema Province and the northern margin of the São Francisco Craton. In: Revista Brasileira de Geociências, v. 22: pp. 472-480

Mendes, V.A., Brito, M. de F. L. (Orgs.). Arapiraca. Sheet SC.24-X-D-V: state of Alagoas scale 1:100.000. Brasília: CPRM; DNPM, 100 p. il. + 2 colour maps. Basic Geological Surveys of Brazil Programme - PLGB (2011)

Mendes, V.A; Brito, M.F.L.; Paiva, IP., 2009. Geology of Brazil Programme-PGB. Arapiraca. Sheet SC.24-X-D. States of Alagoas, Pernambuco and Sergipe.Geological Map. Recife: CPRM, 1 map, colour, 112.37 cm x 69.42 cm. Scale - 1 :250,000. 1999

Nascimento, R.S., Oliveira, E.P., Carvalho, M.J., Mcnaughton, N., 2005. Tectonic evolution of the Canindé Domain, Sergipana Belt, NE Brazil. In: III Symposium on the São Francisco Craton, Salvador, Bahia, proceedings, pp. 239-242.

Oliveira, E.P., Windley, B., Araújo, M.N.C., (2010). The Neoproterozoic Sergipano orogenic belt, NE Brazil: a complete plate tectonic cycle in western Gondwana.Precambrian Res. 181, 64-84.

Oliveira, Elson P.; Windley, Brian F.; Araújo, Mario NC. 2006. Geologic correlation between the Neoproterozoic Sergipano belt (NE Brazil) and the Yaounde' belt (Cameroon, Africa). Journal of African Earth Sciences, v. 44, p. 470-478.

Pettijohn, F.J., Potter, PE., Siever, R., 1972. Sand and Sandstones: New York, Springer-Verlag. 553 pp.

Santos, E.J; Brito Neves, B.B. Borborema Province. 1984 In: Almeida F. F. M.; Hasui y. (eds). The Precambrian of Brazil. São Paulo, Edgar Blucher, p. 123-186.

State Secretariat for the Environment and Water Resources of the State of Alagoas, SEMARH-AL. Monthly precipitation data .   Availablein http://dados.al.gov.br/it/dataset/dados-de-precipitacao-   monthly/resource/0f8aa4fe-48f4-432a-b9df-cb94b323186b Accessed 20/03/2016

Silva Filho, M.A., 2006. Lithogeochemistry and Evolution of the Marancó Domain of the Sergipano System, Northeast Brazil. PhD Thesis, Federal University of Pernambuco. Recife 213p.

Silva Filho, M.A., 1998. Canindé-Marancó volcanic arc and the Sul-Alagoas Belt: orogenic sequences Mesoproterozoic. IX Congr. Bras. Geol., Belo Horizonte, 16.

Silva Filho, M.A., Acioly, A.C.A., Torres, H.H.F., Araújo, R.V., 2003. The Jaramataia Complex in the Context of the Sergipano System. Revista de Geologia 16, 99-110.

Van Schmus, W. R.; Kozuchm.; Brito Neves B. B., 2011. Precambrian History of the Transverse Zone of the Borborema Province, NE Brazil: Insights from SM-ND and U- PB Geochronology. Journal of South American Earth Sciences. Vol. 31 Pages 227-252

Annex A - (Points visited and structural data)

Outcrop	Latitude (S)	Longitude (W)	Description of the outcrop	Foliation (DIR./INT.)	
RS-01	9° 41' 22"	37° 8' 17"	Limestone		
RS-02	9° 41' 46"	37° 9' 10"	Granada-biotite paragnase		
RS-03	9° 43' 25"	37° 10' 10"	Granada-biotite paragnase	N2°W	860°
RS-04	9° 40' 40"	37° 8' 2"	Granada-biotite paragnase	N25°W	33
RS-05	9° 40' 51"	37° 8' 53"	Amphibolite	N33°W	40°
RS-06	9° 40' 41"	37° 9' 40"	Granada-biotite paragnase	N20°W	40°
RS-07	9° 40' 20"	37° 10' 13"	Leucocratic sheet		
RS-08	9° 39' 52"	37° 10' 45"	Amphibolite	N10°W	46°
RS-09	9° 39' 29"	37° 11' 23"	Granada-biotite paragnase		
RS-10	9° 39' 2"	37° 11' 48"	Granada-biotite paragnase	N20°W	40
RS-11	9° 43' 20"	37° 7' 44"	Garnet-biotite schist	N10°W	70°
RS-12	9° 44' 23"	37° 8' 41"	Garnet-biotite schist	N8°W	35°
RS-13	9° 44' 7"	37° 8' 20"	Garnet-biotite schist	N14°W	30°
RS-14	9° 43' 1"	37° 9' 26"	Amphibolite		
RS-15	9° 42' 45"	37° 10' 54"	Amphibolite		
RS-16	9° 39' 45"	37° 7' 24"	Amphibolite	N20°W	35°
RS-17	9° 38' 51"	37° 7' 38"	Granada-biotite paragnase		
RS-18	9° 39' 25"	37° 7' 34"	Amphibolite	N25°W	35°
RS-19	9° 40' 39"	37° 7' 38"	Limestone		
RS-20	9° 43' 24"	37° 7' 42"	Garnet-biotite schist		
RS-21	9° 43' 43"	37° 8' 10"	Quartzite		
RS-22	9° 43' 31"	37° 7' 55"	Shale	N10°S	50°
RS-23	9° 40' 49"	37° 7' 32"	Limestone		
RS-24	9° 44' 43"	37° 9' 55"	Quartzite		
RS-25	9° 43' 30"	37° 8' 22"	Limestone		
RS-26	9° 44' i"	37° 8' 12"	Garnet-biotite schist	N26°W	33°
RS-27	9° 42' 4"	37° 7' 22"	Garnet-biotite schist		

Annex B - Geochemical data for schists and paragneisses

	SCHISTS AND PARAGNEISSES						
	RS-27	RS-09	RS-12	RS-22	RS-20	RS-17	RS-26
SiO_2	60,5	62,14	66,56	66,96	66,64	64,52	69,25
Al_2O_3	16,13	16,1	13,04	14,5	15,51	14,93	11,71
MgO	2,33	2,34	1,27	2,02	1,9	2,43	2,37

MnO	0,06	0,12	0,08	0,13	0,2	0,14	0,11
CaO	3,27	2,5	4,13	1,74	2,09	1,62	2,99
Na2O	2,57	2,58	0,78	0,44	1,32	1,52	1,4
K2O	2,74	2,56	3,59	3,8	1,94	2,88	2,67
TiO	0,93	0,94	0,92	0,9	0,81	1,32	0,77
P O25	0,3	0,17	0,18	0,29	0,68	0,14	0,08
Fe2O3	6,99	8,25	5,81	8,34	8,95	9,86	6,3
PF	2,41	1,89	2,33	1,45	2,54	2,14	1,78
total	98,23	99,59	98,69	100,57	102,58	101,5	99,43
Cr	206	304	272	180	187	191	162
Ba	953	816	516	1258	1285	722	128
Rb	149	108	168	??	165	133	197
Mr	707	470	256	99	967	215	352
Zr	244	248	60	316	292	308	136
Y	40	42	27	55	37	45	41
Nb	8	13	16	16	22	20	12
Ni	35	64	15	76	26	60	18

Annex B2 - Geochemical data of the amphibolites

	ANFIBOLITES				
	RS-O5	RS-16	RS-14	RS-18	RS-15
SiO$_2$	50,11	65,6	56,45	62,98	49,25
Al O$_{23}$	18,58	16,51	13,87	15,26	16,91
MgO	3,42	1,75	6,55	2,41	4,46
MnO	0,14	0,07	0,09	0,07	0,13
CaO	8,82	4,43	7,09	3,96	11,15
Na2O	2,35	2,86	1,05	2,81	2,57
K2O	2,24	2,71	2,13	3,24	0,71
TiO	0,65	0,65	2,18	0,88	1,38
P O$_{25}$	0,04	0,08	0,21	0,21	0,08
Fe2O3	9,84	5,64	11,02	6,34	9,75
PF	2,2	2,14	1,66	2,14	1,82
total	98,39	102,44	102,3	100,3	98,21
Cr	166	177	212	258	131
Ba	756	675	804	728	185
Rb	65	161	162	266	36
Mr	493	303	287	354	341
Zr	64	281	251	324	113
Y	28	36	28	43	18
Nb	2	17	26	13	16
Ni	60	16	16	65	41

Annex B3 (Limestone geochemical data)

	Limestone			
	RS-19	RS-01	RS-23	RS-25
SiO$_2$	2,17	5,57	4,47	3,10
Al O$_{23}$	1,16	0,14	0,00	0,52
MgO	3,51	17,92	20,37	12,41
MnO	0,16	0,03	0,02	0,02
CaO	51,35	33,22	32,29	40,94
Na2O	0,00	0,02	0,10	0,04
K2O	0,00	0,04	0,02	0,01
TiO	0,09	0,02	0,01	0,04
P O$_{25}$	0,03	0,00	0,00	0,03
Fe O T$_{23}$	0,06	0,55	0,21	0,44
PF	40,41	40,55	42,36	40,97
total	98,94	98,06	99,86	98,51
Rb	1	3	2	3
Mr	1213	198	83	103

Printed by Books on Demand GmbH, Norderstedt / Germany